Celestial

Tabitha L. Barnett

Copyright ©2017 Tabitha L Barnett

This publication is protected by copyright law. Please respect the law. No part of this publication can be reproduced, reused, republished, or distributed in any form or by any means, electronic or mechanical, or stored in a database or retrieval system without prior written consent from the artist. The one exception to this policy is that you are permitted to photocopy the original pages in this book to color for your own personal use.

©2017 Tabitha Barnett   facebook.com/tabbystangledart

©2017 Tabitha Barnett   facebook.com/tabbystangledart

©2017 Tabitha Barnett facebook.com/tabbystangledart

©2017 Tabitha Barnett facebook.com/tabbystangledart

©2017 Tabitha Barnett   facebook.com/tabbystangledart

©2017 Tabitha Barnett   facebook.com/tabbystangledart

©2017 Tabitha Barnett   facebook.com/tabbystangledart

©2017 Tabitha Barnett facebook.com/tabbystangledart

©2017 Tabitha Barnett   facebook.com/tabbystangledart

©2017 Tabitha Barnett   facebook.com/tabbystangledart

©2017 Tabitha Barnett	facebook.com/tabbystangledart

©2017 Tabitha Barnett   facebook.com/tabbystangledart

©2017 Tabitha Barnett   facebook.com/tabbystangledart

©2017 Tabitha Barnett facebook.com/tabbystangledart

©2017 Tabitha Barnett   facebook.com/tabbystangledart

©2017 Tabitha Barnett   facebook.com/tabbystangledart

©2017 Tabitha Barnett   facebook.com/tabbystangledart

©2017 Tabitha Barnett facebook.com/tabbystangledart

©2017 Tabitha Barnett facebook.com/tabbystangledart

©2017 Tabitha Barnett   facebook.com/tabbystangledart

©2017 Tabitha Barnett   facebook.com/tabbystangledart

©2017 Tabitha Barnett   facebook.com/tabbystangledart

Color Test Sheet

Join the conversation on facebook:
www.facebook.com/tabbystangledart

If you enjoyed this book, please consider taking a few minutes to leave a review on Amazon.

Please post your colored images online with
#tabbystangledart or #tabbyb so I can find them easily.

Instagram: @tabbystangledart
Twitter: @tabbyleann
www.patreon.com/tabbyb
www.sellfy.com/tabbyb
www.redbubble.com/people/tabbyb
www.amazon.com/author/tabbystangledart
http://tinyurl.com/tabbytube

Made in the USA
Lexington, KY
25 May 2018